History of Paleontology

# History of Paleontology

Thomas H. Huxley
Charles O. Marsh

*History and Civilization Collection*

LM Publishers

# Part 1

# The Rise and Progress of Paleontology[1]

That application of the sciences of biology and geology which is commonly known as paleontology took its origin in the mind of the first person who, finding something like a shell or a bone naturally imbedded in gravel or in rock, indulged in speculations upon the nature of this thing which he had dug out—this "fossil"—and upon the causes which had brought it into such a position. In this rudimentary form, a high antiquity may safely be ascribed to paleontology, inasmuch as we know that, five hundred years before the Christian era, the philosophic doctrines of Xenophanes were influenced by his observations upon the fossil remains exposed in the quarries of Syracuse. From this time forth, not only the philosophers, but the poets, the historians, the geographers of antiquity occasionally refer to fossils; and after the revival of learning lively controversies arose respecting their real nature. But hardly more than two centuries have elapsed since

---

[1] By Thomas H. Huxley

7

this fundamental problem was first exhaustively treated; it was only in the last century that the archæological value of fossils—their importance, I mean, as records of the history of the earth—was fully recognized; the first adequate investigation of the fossil remains of any large group of vertebrated animals is to be found in Cuvier's "Recherches sur les Ossemens Fossiles," completed in 1822; and so modern is stratigraphical paleontology, that its founder, William Smith, lived to receive the just recognition of his services by the award of the first Wollaston medal in 1831.

But, although paleontology is a comparatively youthful scientific specialty, the mass of materials with which it has to deal is already prodigious. In the last fifty years the number of known fossil remains of invertebrated animals has been trebled or quadrupled. The work of interpretation of vertebrate fossils, the foundations of which were so solidly laid by Cuvier, was earned on with wonderful vigor and success by Agassiz in Switzerland, by Von Meyer in Germany, and, last but not least, by Owen in this country, while, in later years, a multitude of workers have labored in the same field. In many groups of the animal kingdom the number of fossil forms already known is as great as that of the existing species. In some cases it is much greater; and there are entire orders of animals of the existence of which we should know nothing except for the evidence afforded by fossil remains. With all this it may be safely assumed that, at the present moment, we are not acquainted with a tithe of the fossils which will sooner or later be discovered. If we may judge by the profusion yielded within the last few years by the Tertiary formations of North America, there seems to be no limit to the multitude of mammalian remains to be expected from that continent, and analogy leads us to expect similar riches in Eastern Asia whenever the Tertiary formations of that region are as carefully explored. Again, we have as yet almost everything to learn respecting the terrestrial population of the Mesozoic epoch—and it seems as if the Western Territories of the United States were about to prove as instructive in regard to this point as they have in respect of Tertiary life. My friend Professor Marsh informs me that, within two years, remains of more than one hundred and sixty distinct individuals of mammals, belonging to twenty species and nine genera, have been found in a space not larger than the floor of a good-sized room; while beds of the same age have yielded three hundred reptiles, varying in size from a length of sixty or eighty feet to the dimensions of a rabbit.

The task which I have set myself to-night is to endeavor to lay before you, as briefly as possible, a sketch of the successive steps by which our present knowledge of the facts of paleontology and of those conclusions from them which are indisputable has been attained; and I beg leave to remind you at the outset that, in attempting to sketch the progress of a branch of knowledge to which innumerable labors have contributed, my business is rather with generalizations than with details. It is my object to mark the epochs of paleontology, not to recount all the events of its history.

That which I just now called the fundamental problem of paleontology, the question which has to be settled before any other can be profitably discussed, is this: What is the nature of fossils? Are they, as the healthy common-sense of the ancient Greeks appears to have led them to assume without hesitation, the remains of animals and plants? Or are they, as was so generally maintained in the fifteenth, sixteenth, and seventeenth centuries, mere figured stones, portions of mineral matter which have assumed the forms of leaves and shells and bones, just as those portions of mineral matter which we call crystals take on the form of regular geometrical solids? Or, again, are they, as others thought, the products of the germs of animals and of the seeds of plants which have lost their way, as it were, in the bowels of the earth, and have achieved only an imperfect and abortive development? It is easy to sneer at our ancestors for being disposed to reject the first in favor of one or other of the last two hypotheses; but it is much more profitable to try to discover why they, who were really not one whit less sensible persons than our excellent selves, should have been led to entertain views which strike us as absurd. The belief in what is erroneously called spontaneous generation—that is to say, in the development of living matter out of mineral matter, apart from the agency of pre-existing living matter, as an ordinary occurrence at the present day—which is still held by some of us, was universally accepted as an obvious truth by them. They could point to the arborescent forms assumed by hoar-frost and by sundry metallic minerals as evidence of the existence in nature of a "plastic force" competent to enable inorganic matter to assume the form of organized bodies. Then, as everyone who is familiar with fossils knows, they present innumerable gradations, from shells and bones which exactly resemble the recent objects, to masses of mere stone which, however accurately they repeat the outward form of the organic body, have nothing else in common with it; and, thence, to mere traces and faint impressions in the continuous substance of the rock. What we now know to be the results of the chemical changes which take place in the course of fossilization, by which mineral is substituted for organic substance, might, in the absence of such knowledge, be fairly interpreted as the expression of a process of development in the opposite direction—from the mineral to the organic. Moreover, in an age when it would have seemed the most absurd of paradoxes to suggest that the general level of the sea is constant, while that of the solid land fluctuates up and down through thousands of feet in a secular ground-swell, it may well have appeared far less hazardous to conceive that fossils are sports of Nature than to accept the necessary alternative, that all the inland regions and highlands, in the rocks of which marine shells had been found, had once been covered by the ocean. It is not so surprising, therefore, as it may at first seem, that, although such men as Leonardo da Vinci and Bernard Palissy took just views of the nature of fossils, the opinion of the majority of their contemporaries set strongly the other way; nor even that error maintained itself long after the scientific grounds of the true interpretation of fossils had been stated, in a manner that left nothing to be desired, in the latter half of the seventeenth century. The person who rendered this good service to paleontology was Nicholas Steno, professor of anatomy in Florence, though a Dane by birth. Collectors of fossils at that day were familiar with certain bodies termed "glossopetræ," and speculation was rife as to their nature. In the first half of the seventeenth century, Fabio Colonna had tried to convince his colleagues of the famous Accademia dei Lincei that the glossopetræ were merely fossil sharks' teeth, but his arguments made no impression. Fifty years later Steno reopened the question, and, by dissecting the head of a shark and pointing out the very exact correspondence of its teeth with the glossopetræ, left no rational doubt as to the origin of the latter. Thus far, the work of Steno went little further than that of Colonna, but it fortunately occurred to him to think out the whole subject of the interpretation of fossils, and the result of his meditations was the publication, in 1669, of a little treatise with the very quaint title of "De Solido intra Solidum naturaliter contento." The general course of Steno's argument may be stated in a few words. Fossils are solid bodies which by some natural process have come to be contained within other solid bodies—namely, the rocks in which they are imbedded; and the fundamental problem of paleontology, stated generally, is this: "Given a body endowed with a certain shape and produced in accordance with natural laws, to find in that body itself the evidence of the place and manner of its production." The only way of solving this problem is by the application of the axiom that "like effects imply like causes," or as Steno puts it, in reference to this particular case, that "bodies which are altogether similar have been produced in the same way." Hence, since the glossopetræ are altogether similar to sharks' teeth, they must have been produced by shark-like fishes; and since many fossil shells correspond, down to the minutest details of structure, with the shells of existing marine or fresh-water animals, they must have been produced by similar animals; and the like reasoning is applied by Steno to the fossil bones of vertebrated animals, whether aquatic or terrestrial. To the obvious objection that many fossils are not altogether similar to their living analogues, differing in substance while agreeing in form, or being mere hollows or impressions, the surfaces of which are figured in the same way as those of animal or vegetable organisms, Steno replies by pointing out the changes which take place in organic remains imbedded in the earth, and how their solid substance may be dissolved away entirely, or replaced by mineral matter, until nothing is left of the original but a cast, an impression, or a mere trace of its contours. The principles of investigation thus excellently stated and illustrated by Steno in 1669, are those which have, consciously or unconsciously, guided the researches of paleontologists ever since. Even that feat of paleontology which has so powerfully impressed the popular imagination, the reconstruction of an extinct animal from a tooth or a bone, is based upon the simplest imaginable application of the logic of Steno. A moment's consideration will show, in fact, that Steno's conclusion that the glossopetræ are sharks' teeth implies the reconstruction of an animal from its tooth. It is equivalent to the assertion that the animal of which the glossopetræ are relics, had the form and organization of a shark; that it had a skull, a vertebral column, and limbs similar to those which are characteristic of this group of fishes; that its heart, gills, and intestines presented the peculiarities which those of all sharks exhibit; nay, even that any hard parts which its integument contained were of a totally different character from the scales of ordinary fishes. These conclusions are as certain as any based upon probable reasonings can be. And they are so, simply because a very large experience justifies us in believing that teeth of this particular form and structure are invariably associated with the peculiar organization of sharks, and are never found in connection with other organisms. Why this should be we are not at present in a position even to imagine; we must take the fact as an empirical law of animal morphology, the reason of which may possibly be one day found in the history of the evolution of the shark tribe, but for which it is hopeless to seek for an explanation in ordinary physiological reasonings. Every one practically acquainted with paleontology is aware that it is not every tooth nor every bone which enables us to form a judgment of the character of the animal to which it belonged, and that it is possible to possess many teeth, and even a large portion of the skeleton of an extinct animal, and yet be unable to reconstruct its skull or its limbs. It is only when the tooth or bone presents peculiarities which Ave know by previous experience to be characteristic of a certain group that we can safely predict that the fossil belonged to an animal of the same group. Anyone who finds a cow's grinder may be perfectly sure that it belonged to an animal which had two complete toes on each foot, and ruminated; anyone who finds a horse's grinder may be as sure that it had one complete toe on each foot and did not ruminate; but, if ruminants and horses were extinct animals of which nothing but the grinders had ever been discovered, no amount of physiological reasoning could have enabled us to reconstruct either animal, still less to have divined the wide differences between the two. Cuvier, in the "Discours sur les Revolutions de la Surface du Globe," strangely credits himself, and has ever since been credited by others, with the invention of a new method of paleontological research. But if you will turn to the "Recherches sur les Ossemens Fossiles," and watch Cuvier, not speculating but working, you will find that his method is neither more nor less than that of Steno. If he was able to make his famous prophecy from the jaw which lay upon the surface of a block of stone to the pelvis of the same animal which lay hidden in it, it was not because either he, or anybody else, knew, or knows, why a certain form of jaw is, as a rule, constantly accompanied by the presence of marsupial bones— but simply because experience has shown that these two structures are co-ordinated.

The settlement of the nature of fossils led at once to the next advance of paleontology—viz., its application to the deciphering of the history of the earth. When it was admitted that fossils are remains of animals and plants, it followed that, in so far as they resemble terrestrial or fresh-water animals and plants, they are evidences of the existence of land or fresh water, and, in so far as they resemble marine organisms, they are evidences of the existence of the sea at the time at which they were parts of actually living animals and plants. Moreover, in the absence of evidence to the contrary, it must be admitted that the terrestrial or the marine organisms implied the existence of land or sea at the place in which they were found while they were yet living. In fact, such conclusions were immediately drawn by everybody, from the time of Xenophanes downward, who believed that fossils were really organic remains. Steno discusses their value as evidence of repeated alteration of marine and terrestrial conditions upon the soil of Tuscany in a manner worthy of a modern geologist. The speculations of De Maillet in the beginning of the eighteenth century turn upon fossils, and Buffon follows him very closely in those two remarkable works, the "Théorie de la Terre" and the "Époques de la Nature," with which he commenced and ended his career as a naturalist.

The opening sentences of the "Époques de la Nature" show us how fully Buffon recognized the analogy of geological with archaeological inquiries.

"As in civil history we consult deeds, seek for coins, or decipher antique inscriptions in order to determine the epochs of human revolutions and fix the date of moral events, so, in natural history, we must search the archives of the world, recover old monuments from the bowels of the earth, collect their fragmentary remains, and gather into one body of evidence all the signs of physical change which may enable us to look back upon the different ages of nature. It is our only means of fixing some points in the immensity of space, and of setting a certain number of way-marks along the eternal path of time."

Buffon enumerates five classes of these monuments of the past history of the earth, and they are all facts of paleontology. In the first place, he says, shells and other marine productions are found all over the surface and in the interior of the dry land; and all calcareous rocks are made up of their remains. Secondly, a great many of these shells which are found in Europe are not now to be met with in the adjacent seas; and, in the slates and other deep-seated deposits, there are remains of fishes and of plants of which no species now exist in our latitudes, and which are either extinct or exist only in more northern climates. Thirdly, in Siberia and in other northern regions of Europe and of Asia, bones and teeth of elephants, rhinoceroses, and hippopotamuses occur in such numbers that these animals must once have lived and multiplied in those regions, although at the present day they are confined to southern climates. The deposits in which these remains are found are superficial, while those which contain shells and other marine remains lie much deeper. Fourthly, tusks and bones of elephants and hippopotamuses are found not only in the northern regions of the Old World, but also in those of the New World, although, at present, neither elephants nor hippopotamuses occur in America. Fifthly, in the middle of the continents, in regions most remote from the sea, we find an infinite number of shells, of which the most part belong to animals of those kinds which still exist in southern seas, but of which many others have no living analogues; so that these species appear to be lost, destroyed by some unknown cause. It is needless to inquire how far these statements are strictly accurate; they are sufficiently so to justify Buffon's conclusions that the dry land was once beneath the sea; that the formation of the fossiliferous rocks must have occupied a vastly greater lapse of time than that traditionally ascribed to the age of the earth; that fossil remains indicate different climatal conditions to have obtained in former times, and especially that the polar regions were once warmer; that many species of animals and plants have become extinct; and that geological change has had something to do with geographical distribution.

But these propositions almost constitute the framework of paleontology. In order to complete it but one addition was needed, and that was made, in the last years of the eighteenth century, by William Smith, whose work comes so near our own times that many living men may have been personally acquainted with him. This modest land-surveyor, whose business took him into many parts of England, profited by the peculiarly favorable conditions offered by the arrangement of our secondary strata to make a careful examination and comparison of their fossil contents at different points of the large area over which they extend. The result of his accurate and widely extended observations was to establish the important truth that each stratum contained certain fossils which are peculiar to it; and that the order in which the strata, characterized by these fossils, are superimposed one upon the other is always the same. This most important generalization was rapidly verified and extended to all parts of the world accessible to geologists; and now it rests upon such an immense mass of observations as to be one of the best established truths of natural science. To the geologist this discovery was of infinite importance, as it enabled him to identify rocks of the same relative age, however their continuity might be interrupted or their composition altered. But to the biologist it had a still deeper meaning, for it demonstrated that, throughout the prodigious duration of time registered by the fossiliferous rocks, the living population of the earth had undergone continual changes, not merely by the extinction of a certain number of the species which at first existed, but by the continual generation of new species, and the no less constant extinction of old ones.

Thus, the broad outlines of paleontology, in so far as it is the common property of both the geologist and the biologist, were marked out at the close of the last century. In tracing its subsequent progress I must confine myself to the province of biology, and indeed to the influence of paleontology upon zoölogical morphology. And I accept this limitation the more willingly as the no less important topic of the bearing of geology and of paleontology upon distribution has been luminously treated in the address of the President of the Geographical Section.

The succession of the species of animals and plants in time being established, the first question which the zoölogist or the botanist had to ask himself was, "What is the relation of these successive species one to another?" And it is a curious circumstance that the most important event in the history of paleontology which immediately succeeded William Smith's generalization was a discovery which, could it have been rightly appreciated at the time, would have gone far toward suggesting the answer, which was in fact delayed for more than half a century. I refer to Cuvier's investigation of the mammalian fossils yielded by the quarries in the older Tertiary rocks of Montmartre, among the chief results of which was the bringing to light of two genera of extinct hoofed quadrupeds, the *Anoplotherium* and the *Palæotherium*. The rich materials at Cuvier's disposition enabled him to obtain a full knowledge of the osteology and of the dentition of these two forms, and consequently to compare their structure critically with that of existing hoofed animals. The effect of this comparison was to prove that the *Anoplotherium*, though it presented many points of resemblance with the pigs on the one hand, and with the ruminants on the other, differed from both to such an extent that it could find a place in neither group. In fact, it held, in some respects, an intermediate position, tending to bridge over the interval between these two groups, which in the existing fauna are so distinct. In the same way, the Palæotherium tended to connect forms so different as the tapir, the rhinoceros, and the horse. Subsequent investigations have brought to light a variety of facts of the same order, the most curious and striking of which are those which prove the existence, in the Mesozoic epoch, of a series of

25

forms intermediate between birds and reptiles—two classes of vertebrate animals which at present appear to be more widely separated than any others. Yet the interval between them is completely filled, in the mesozoic fauna, by birds which have reptilian characters on the one side, and reptiles which have ornithic characters on the other. So, again, while the group of fishes termed ganoids is at the present time so distinct from that of the dipnoi or mud-fishes that they have been reckoned as distinct orders, the Devonian strata present us with forms of which it is impossible to say with certainty whether they are dipnoi or whether they are ganoids.

Agassiz's long and elaborate researches upon fossil fishes, published between 1833 and 1842, led him to suggest the existence of another kind of relation between ancient and modern forms of life. He observed that the oldest fishes presented many characters which recall the embryonic conditions of existing fishes; and that, not only among fishes, but in several groups of the invertebrata which have a long paleontological history, the latest forms are more modified, more specialized, than the earlier.

The fact that the dentition of the older tertiary ungulate and carnivorous mammals is always complete, noticed by Professor Owen, illustrated the same generalization.

Another no less suggestive observation was made by Mr. Darwin, whose personal investigations during the voyage of the Beagle led him to remark upon the singular fact that the fauna which immediately precedes that at present existing in any geographical province of distribution presents the same peculiarities as its successor. Thus, in South America and in Australia, the later tertiary or quaternary fossils show that the fauna which immediately preceded that of the present day was, in the one case, as much characterized by edentates and in the other by marsupials as it is now, although the species of the older are largely different from those of the newer fauna.

However clearly these indications might point in one direction, the question of the exact relation of the successive forms of animal and vegetable life could be satisfactorily settled only in one way— namely, by comparing, stage by stage, the series of forms presented by one and the same type throughout a long space of time. Within the last few years this has been done fully in the case of the horse, less completely in the case of the other principal types of the ungulata and of the carnivora, and all these investigations tend to one general result—namely, that in any given series the successive members of that series present a gradually increasing specialization of structure. That is to say, if any such mammal at present existing has specially modified and reduced limbs or dentition and complicated brain, its predecessors in time show less and less modification and reduction in limbs and teeth and a less highly developed brain. The labors of Gaudry, Marsh, and Cope furnish abundant illustrations of this law from the marvelous fossil wealth of Pikermi, and the vast uninterrupted series of tertiary rocks in the Territories of North America.

I will now sum up the results of this sketch of the rise and progress of paleontology. The whole fabric of paleontology is based upon two propositions: the first is, that fossils are the remains of animals and plants; and the second is, that the stratified rocks in which they are found are sedimentary deposits; and each of these propositions is founded upon the same axiom that like effects imply like causes. If there is any cause competent to produce a fossil stem, or shell, or bone, except a living being, then paleontology has no foundation; if the stratification of the rocks is not the effect of such causes as at present produce stratification, we have no means of judging of the duration of past time, or of the order in which the forms of life have succeeded one another. But, if these two propositions are granted, there is no escape, as it appears to me, from three very important conclusions. The first is, that living matter has existed upon the earth for a vast length of time, certainly for millions of years. The second is that, during this lapse of time, the forms of living matter have undergone repeated changes, the effect of which has been that the animal and vegetable population at any period of the earth's history contains some species which did not exist at some antecedent period, and others which ceased to exist at some subsequent period. The third is that, in the case of many groups of mammals and some of reptiles, in which one type can be followed through a considerable extent of geological time, the series of different forms by which the type is represented at successive intervals of this time is exactly such as it would be if they had been produced by the gradual modification of the earliest form of the series. These are facts of the history of the earth guaranteed by as good evidence as any facts in civil history.

Hitherto I have kept carefully clear of all the hypotheses to which men have at various times endeavored to fit the facts of paleontology, or by which they have endeavored to connect as many of these facts as they happened to be acquainted with. I do not think it would be a profitable employment of our time to discuss conceptions which doubtless have had their justification and even their use, but which are now obviously incompatible with the well-ascertained truths of paleontology. At present these truths leave room for only two hypotheses. The first is that, in the course of the history of the earth, innumerable species of animals and plants have come into existence, independently of one another, innumerable times. This, of course, implies either that spontaneous generation on the most astounding scale, and of animals such as horses and elephants, has been going on, as a natural process, through all the time recorded by the fossiliferous rocks; or it necessitates the belief in innumerable acts of creation repeated innumerable times. The other hypothesis is, that the successive species of animals and plants have arisen, the later by the gradual modification of the earlier. This is the hypothesis of evolution; and the paleontological discoveries of the last decade are so completely in accordance with the requirements of this hypothesis that, if it had not existed, the paleontologist would have had to invent it.

I have always had a certain horror of presuming to set a limit upon the possibilities of things. Therefore, I will not venture to say that it is impossible that the multitudinous species of animals and plants may have been produced one separately from the other by spontaneous generation, nor that it is impossible that they should have been independently originated by an endless succession of miraculous creative acts. But I must confess that both these hypotheses strike me as so astoundingly improbable, so devoid of a shred of either scientific or traditional support, that even if there were no other evidence than that of paleontology in its favor, I should feel compelled to adopt the hypothesis of evolution. Happily, the future of paleontology is independent of all hypothetical considerations. Fifty years hence, whoever undertakes to record the progress of paleontology will note the present time as the epoch in which the law of succession of the forms of the higher animals was determined by the observation of paleontological facts. He will point out that, just as Steno and as Cuvier were enabled from their knowledge of the empirical laws of coexistence of the parts of animals to conclude from a part to the whole, so the knowledge of the law of succession of forms empowered their successors to conclude, from one or two terms of such a succession, to the whole series, and thus to divine the existence of forms of life, of which, perhaps, no trace remains, at epochs of inconceivable remoteness in the past.

## Part 2

# History and Methods of Paleontological Discovery[2]

I

In the rapid progress of knowledge, we are constantly brought face to face with the question. What is life? The answer is not yet, but a thousand earnest seekers after truth seem to be slowly approaching a solution. This question gives a new interest to every department of science that relates to life in any form, and the history of life offers a most suggestive field for research. One line of investigation lies through embryology, and here the advance is most encouraging. Another promising path leads back through the life-history of the globe, and in this direction we may hope for increasing light, as a reward for patient work.

The plants and animals now living on the earth interest alike the savage and the *savant,* and hence have been carefully observed in every age of human

---

[2] By Charles O. Marsh.

history. The life of the remote past, however, is preserved only in scanty records, buried in the earth, and therefore readily escapes attention. For these reasons, the study of ancient life is one of the latest of modern sciences, and among the most difficult. In view of the great advances which this department of knowledge has made within the last decade, especially in this country, I have thought it fitting to the present occasion to review briefly its development, and have chosen for my subject this evening "The History and Methods of Paleontological Discovery."

In the short time now at my command, I can only attempt to present a rapid sketch of the principal steps in the progress of this science. The literature of the subject, especially in connection with the discussions it provoked, is voluminous, and an outline of the history itself must suffice for my present purpose.

In looking over the records of paleontology, its history may conveniently be divided into four periods, well marked by prominent features, but, like all stages of intellectual growth, without definite boundaries.

*The first period,* dating back to the time when men first noticed fossil remains in the rocks, and queried as to their nature, is of special historic interest. The most prominent characteristic of this period was, a long and bitter contest as to the *nature of fossil remains.* Were they mere "sports of Nature," or had they once been endowed with life? Simple as this problem now seems, centuries passed before the wise men of that time were agreed upon its solution.

Sea-shells in the solid rock on the tops of mountains early attracted the attention of the ancients, and the learned men among them seem to have appreciated in some instances their true character, and given rational explanations of their presence.

The philosopher Zenophanes, of Colophon, who lived about 500 B. C., mentions the remains of fishes and other animals in the stone quarries near Syracuse, the impression of an anchovy in the rock of Paros, and various marine fossils at other places. His conclusion from these facts was, that the surface of the earth had once been in a soft condition at the bottom of the sea; and thus the objects mentioned were entombed. Herodotus, half a century later, speaks of marine shells on the hills of Egypt and over the Libyan Desert, and he inferred therefrom that the sea had once covered that whole region. Empedocles, of Agrigentum (450 B. C.), believed that the many hippopotamus-bones found in Sicily were remains of human giants, in comparison with which the present race were as children. Here, he thought, was a battle-field between the gods and the Titans, and the bones belonged to the slain. Pythagoras (582 b. c.) had already anticipated one conclusion of modern geology, if the following statement, attributed to him by Ovid, was his own:

"Vidi ego quod fuerat solidissima tellus,
Esse fretum: vidi factas ex æquore terras;
Et procul a pelago conchæ Jacuere marinæ."

Aristotle (384-322 B. C.) was not only aware of the existence of fossils in the rocks, but has also placed on record sagacious views as to the changes in the earth's surface necessary to account for them. In the second book of his "Meteorics," he says: "The changes of the earth are so slow in comparison to the duration of our lives, that they are overlooked; and the migrations of people after great catastrophes and their removal to other regions, cause the event to be forgotten." Again, in the same work, he says: "As time never fails, and the universe is eternal, neither the Tanais nor the Nile can have flowed forever. The places where they rise were once dry, and there is a limit to their operations: but there is none to time. So of all other rivers; they spring up, and they perish; and the sea also continually deserts some lands and invades others. The same tracts, therefore, of the earth are not, some always sea, and others always continents, but everything changes in the course of time."

Aristotle's views on the subject of spontaneous generation were less sound, and his doctrines on this subject exerted a powerful influence for the succeeding twenty centuries. In the long discussion that followed concerning the nature of fossil remains, Aristotle's views were paramount. He believed that animals could originate from moist earth or the slime of rivers, and this seemed to the people of that period a much simpler way of accounting for the remains of animals in the rocks than the marvelous changes of sea and land otherwise required to explain their presence. Aristotle's opinion was in accordance with the Biblical account of the creation of man out of the dust of the earth, and hence more readily obtained credence.

Theophrastus, a pupil of Aristotle, alludes to fossil fishes found near Heraclea, in Pontus, and in Paphlagonia, and says, "They were either developed from fish-spawn left behind in the earth, or gone astray from rivers or the sea into cavities of the earth, where they had become petrified." In treating of fossil ivory and bones, the same writer supposed them to be produced by a certain plastic virtue latent in the earth. To this same cause, as we shall see, many later authors attributed the origin of all fossil remains.

Previous to this, Anaximander, the Miletian philosopher, who was born about 610 years before Christ, had expressed essentially the same view. According to both Plutarch and Censorinus, Anaximander taught that fishes, or animals very like fishes, sprang from heated water and earth, and from these animals came the human race; a statement which can hardly be considered as anticipating the modern idea of evolution, as some authors have imagined.

The Romans added but little to the knowledge possessed by the Greeks in regard to fossil remains. Pliny (23-79 A. D.), however, seems to have examined such objects with interest, and in his renowned work on natural history gave names to several forms. He doubtless borrowed largely from Theophrastus, who wrote about three hundred years before. Among the objects named by Pliny were: "*Bucardia,* like to an ox's heart"; *Brontia,* resembling the head of a tortoise, supposed to fall in thunderstorms"; "*Glossoptra,* similar to a human tongue, which does not grow in the earth, but falls from heaven while the moon is eclipsed"; "the *Horn of Ammon,* possessing, with a golden color, the figure of a ram's horn"; *Ceraunia* and *Ombria,* supposed to be thunderbolts"; "*Ostracites,* resembling the oyster shell"; "*Spongites,* having the form of sponge"; *Phycites,* similar to sea-weed or rushes." He also mentions stones resembling the teeth of hippopotamus; and says that Theophrastus speaks of fossil ivory, both black and white, of bones born in the earth, and of stones bearing the figure of bones.

Tertullian (160 A. D.) mentions instances of the remains of sea animals on the mountains, far from the sea, but uses them as a proof of the general deluge recorded in Scripture.

During the next thirteen or fourteen centuries, fossil remains of animals and plants seem to have attracted so little attention, that few references are made to them by the writers of this period. During these ages of darkness, all departments of knowledge suffered alike, and feeble repetitions of ideas derived from the ancients seem to have been about the only contributions of that period to natural science.

Albert the Great (1205-'80 A. D.),the most learned man of his time, mentions that a branch of a tree was found, on which was a bird's nest containing birds, the whole being solid stone. He accounted for this strange phenomenon by the *vis formativa* of Aristotle, an occult force, which, according to the prevalent notions of the time, was capable of forming most of the extraordinary objects discovered in the earth.

Alexander *ab Alexandro,* of Naples, states that he saw, in the mountains of Calabria, a considerable distance from the sea, a variegated hard marble, in which many sea-shells but little changed were heaped, forming one mass with the marble.

With the beginning of the sixteenth century, a great impetus was given to the investigation of organic fossils, especially in Italy, where this study really began. The discovery of fossil shells, which abound in this region, now attracted great attention, and a fierce discussion soon arose as to the true nature of these and other remains. The ideas of Aristotle in regard to spontaneous generation, and especially his view of the hidden forces of the earth, which he claimed had power to produce such remains, now for the first time were seriously questioned, although it was not till nearly two centuries later that these doctrines lost their dominant influence.

Leonardo da Vinci, the renowned painter and philosopher, who was born in 1452, strongly opposed the commonly accepted opinions as to the origin of organized fossils. He claimed that the fossil shells under discussion were what they

41

seemed, and had once lived at the bottom of the sea. "You tell me," he says, "that Nature and the influence of the stars have formed these shells in the mountains; then show me a place in the mountains where the stars at the present day make shelly forms of different ages, and of different species in the same place." Again, he says, "In what manner can such a cause account for the petrifactions in the same place of various leaves, sea-weeds, and marine crabs?"

In 1517, excavations in the vicinity of Verona brought to light many curious petrifactions, which led to much speculation as to their nature and origin. Among the various authors who wrote on this subject was Fracastoro, who declared that the fossils once belonged to living animals, which had lived and multiplied where found. He ridiculed the prevailing ideas that the plastic force of the ancients could fashion stones into organic forms. Some writers claimed that these shells had been left by Noah's flood, but against this idea Fracastoro offered a mass of evidence, which would now seem conclusive, but which then only aroused bitter hostility. That inundation, he said, was too transient; it consisted mainly of fresh water; and, if it had transported shells to great distances, must have scattered them over the surface, not buried them in the interior of mountains.

Conrad Gesner (1516-'65), whose history of animals has been considered the basis of modern zoölogy, published at Zurich, in 1565, a small but important work entitled "De omni rerum fossiliura genere." It contained a catalogue of the collection of fossils made by John Kentmann. This is the oldest catalogue of fossils with which I am acquainted.

George Agricola (1494-1555) was, according to Cuvier, the first mineralogist who appeared after the revival of learning in Europe. In his great work, "De Re Metallica," published in 1546, he mentions various fossil remains, and says they were produced by a certain "*materia pinguis,*" or fatty matter, set in fermentation by heat. Some years later, Bauhin published a descriptive catalogue of the fossils he had collected in the neighborhood of Boll, in Wurtemberg.

Andrew Mattioli, a distinguished botanist, adopted Agricola's notion as to the origin of organized fossils, but admitted that shells and bones might be turned into stone by being permeated by a "lapidifying juice." Falloppio, the eminent professor of anatomy at Padua, believed that fossil shells were generated by fermentation where they were found; and that the tusks of elephants, dug up near Apulia, were merely earthy concretions. Mercati, in 1574, published figures of the fossil shells preserved in the Museum of the Vatican, but expressed the opinion that they were only stones, and owed their peculiar shapes to the heavenly bodies. Olivi, of Cremona, described the fossils in the Museum at Verona, and considered them all "sports of nature."

Palissy, a French author, in 1580, opposed these views, and is said to have been the first to assert in Paris that fossil shells and fishes had once belonged to marine animals. Fabio Colonna appears to have first pointed out that some of the fossil shells found in Italy were marine and some terrestrial.

Another peculiar theory discussed in the sixteenth century deserves mention. This was the vegetation theory, especially advocated by Tournefort and Camerarius, both eminent as botanists. These writers believed that the seeds of minerals and fossils were diffused throughout the sea and the earth, and were developed into their peculiar forms by the regular increment of their particles, similar to the formation of crystals. "How could the *Cornu Ammonis,*" Tournefort asked, which is constantly in the figure of a volute, be formed without a seed containing the same structure in the small as in the larger forms? "Who molded it so artfully, and where are the molds?" The stalactites which formed in caverns in various parts of the world were also supposed to be proofs of this vegetative growth.

Still another theory has been held at various times, and is not yet entirely forgotten, namely: that the Creator made fossil animals and plants just as they are found in the rocks, in pursuance of a plan beyond our comprehension. This theory has never prevailed among those familiar with scientific facts, and hence needs here no further consideration.

An interest in fossil remains arose in England later than on the Continent; but when attention was directed to them, the first opinions as to their origin were not less fanciful and erroneous than those to which we have already referred. Dr. Plot, in his "Natural History of Oxfordshire," published in 1677, considered the origin of fossil shells and fishes to be due to a "plastic virtue, latent in the earth," as Theophrastus had suggested long before. Lhwyd, in his "Lithophylacii Britannici Ichnographia," published in Oxford in 1699, gives a catalogue of English fossils contained in the Ashmolean Museum. He opposed the *vis plastica* theory, and expressed the opinion that the spawn of fishes and other marine animals had been raised with the vapors from the sea, conveyed inland by clouds, and deposited by rain, had permeated into the interior of the earth, and thus produced the fossil remains we find in the rocks. About this time several important works were published in England by Dr. Martin Lister, which did much to diffuse a true knowledge of fossil remains. He gave figures of recent shells side by side with some of the fossil forms, so that the resemblance became at once apparent. The fossil species of shells he called "turbinated and bivalve stones," and adds, "either these were terriginous, or, if otherwise, the animals which they so exactly represent have become extinct."

During the seventeenth century there was a considerable advance in the study of fossil remains. The discussions in regard to the nature and origin of these objects had called attention to them, and many collections were now made, especially in Italy, and also in Germany, where a strong interest in this subject had been aroused. Catalogues of these collections were not unfrequently published, and some of them were illustrated with such accurate figures, that many of the species can now be readily recognized. In this century, too, an important step in advance was made by the collection and description of fossils from particular localities and regions, in distinction from general collections of curiosities.

Casper Schwenkfeld, in 1600, published a catalogue of the fossils discovered in Silesia; in 1622 a detailed description of the renowned Museum of Calceolarius, of Verona, appeared; and in 1642 a catalogue of Besler's collection. Wormius's catalogue was published in 1652, Spener's in 1663, and Septala's in 1666. A description of the Museum of the King of Denmark was issued in 1669, Cottorp's catalogue in 1674, and that of the renowned Kirscher in 1678, Dr. Grew gave an account in 1687 of the specimens in the Museum of Gresham's College in England; and in 1695 Petiver, of London, published a catalogue of his very extensive collection. A catalogue by Fred. Lauchmund, on the fossils of Hildesheim, appeared in 1669, and the fossils of Switzerland were described by John Jacob Wagner in 1689. Among similar works were the dissertations of Gyer, at Frankfort, and Albertus, at Leipsic.

Steno, a Dane, who had been Professor of Anatomy at Padua, published in 1669 one of the most important works of this period. He entered earnestly into the controversy as to the origin of fossil remains, and by dissecting a shark from the Mediterranean, proved that its teeth were identical with some found fossil in Tuscany, He also compared the fossil shells found in Italy with existing species, and pointed out their resemblance. In the same work, Steno expressed some very important views in regard to the different kinds of strata, and their origin, and first placed on record the important fact that the oldest rocks contain no fossils.

Scilla, the Sicilian painter, published in 1670 a work on the fossils of Calabria, well illustrated. He is very severe against those who doubted the organic origin of fossils, but is inclined to consider them relics of the Mosaic deluge.

Another instance of the power of the *lusus naturæ* theory, even at the close of the seventeenth century, deserves mention. In the year 1696 the skeleton of a fossil elephant was dug up at Tonna, near Gotha, in Germany, and was described by William Ernest Tentzel, a teacher in the Gotha Gymnasium. He declared the bones to be the remains of an animal that had lived long before. The medical faculty in Gotha, however, considered the subject, and decided officially that this specimen was only a freak of Nature.

Besides the authors I have mentioned, there were many others who wrote about fossil remains before the close of the seventeenth century, and took part in the general discussion as to their nature and origin. During the progress of this controversy the most fantastic theories were broached and stoutly defended, and, although refuted from time to time by a few clear-headed men, continually sprang up anew, in the same or modified forms. The influence of Aristotle's views of equivocal generation, and especially the scholastic tendency to disputation, so prevalent during the middle ages, had contributed largely to the retardation of progress, and yet a real advance in knowledge had been made. The long contest in regard to the nature of fossil remains was essentially over, for the more intelligent opinion at the time now acknowledged that these objects were not mere "sports of Nature," but had once been endowed with life. At this point, therefore, the first period in the history of paleontology, as I have indicated it, may appropriately end.

It is true that, later still, the old exploded errors about the plastic force and fermentation were from time to time revived, as they have been almost to the present day; but learned men, with few exceptions, no longer seriously questioned that fossils were real organisms, as the ancients had once believed. The many collections of fossils that had been brought together, and the illustrated works that had been published about them, were a foundation for greater progress, and, with the eighteenth century, the second period in the history of paleontology began.

The main characteristic of this period was the general belief that *fossil remains were deposited by the Mosaic deluge*. We have seen that this view had already been advanced, but it was not till the beginning of the eighteenth century that it became the prevailing view. This doctrine was strongly opposed by some courageous men, and the discussion on the subject soon became even more bitter than the previous one, as to the nature of fossils.

In this diluvial discussion theologians and laymen alike took part. For nearly a century the former had it all their own way, for the general public, then as now, believed what they were taught. Noah's flood was thought to have been universal, and was the only general catastrophe of which the people of that day had any knowledge or conception.

The scholars among them were of course familiar with the accounts of Deucalion and his ark, in a previous deluge, as we are to-day with similar traditions held by various races of men. The firm belief that the earth and all it contains was created in six days; that all life on the globe was destroyed by the deluge, except alone what Noah saved; and that the earth and its inhabitants were to be destroyed by fire, was the foundation on which all knowledge of the earth was based. With such fixed opinions, the fossil remains of animals and plants were naturally regarded as relics left by the flood described in Holy Writ. The dominant nature of this belief is seen in nearly all the literature in regard to fossils published at this time, and some of the works which then appeared have become famous on this account.

In 1710 David Büttner published a volume entitled "Rudera Diluvii Testes." He strongly opposed Lhwyd's explanation of the origin of fossils, and referred these objects directly to the flood. The most renowned work, however, of this time, was published at Zurich in 1726, by Scheuchzer, a physician and naturalist, and professor in the University of Altorf. It bore the title "Homo Diluvii Testis." The specimen upon which this work was based was found at Oeningen, and was regarded as the skeleton of a child destroyed by the deluge. The author recognized in this remarkable fossil, not merely the skeleton, but also portions of the muscles, the liver, and the brain. The same author was fortunate enough to discover, subsequently, near Altorf, two fossil vertebræ, which he at once referred to that "accursed race destroyed by the flood!" These, also, he carefully described and figured in his "Physica Sacra," published at Ulm in 1731. Engravings of both were subsequently given in the "Copper-Bible." Cuvier afterward examined these interesting relics, and pronounced the skeleton of the supposed child to be the remains of a gigantic salamander, and the two vertebræ to be those of an ichthyosaurus!

Another famous book appeared in Germany in the same year in which Scheuchzer's first volume was published. The author was John Bartholomew Adam Beringer, professor at the University of Würzburg, and his great work indirectly had an important influence upon the investigation of fossil remains. The history of the work is instructive, if only as an indication of the state of knowledge at that date. Professor Beringer, in accordance with views of his time, had taught his pupils that fossil remains, or "figured stones," as they were called, were mere "sports of Nature." Some of his fun-loving students reasoned among themselves, "If Nature can make figured stones in sport, why cannot we?" Accordingly, from the soft limestone in the neighboring hills, they carved out figures of marvelous and fantastic forms, and buried them at the localities where the learned Professor was accustomed to dig for his fossil treasures. His delight at the discovery of these strange forms encouraged further production, and taxed the ingenuity of these youthful imitators of Nature's secret processes. At last Beringer had a large and unique collection of forms, new to him and to science, which he determined to publish to the world. After long and patient study his work appeared, in Latin, dedicated to the reigning prince of the country, and illustrated with twenty-one folio plates. Soon after the book was published the deception practiced upon the credulous Professor became known; and, in place of the glory he expected from his great undertaking, he received only ridicule and disgrace. He at once endeavored to repurchase and destroy the volumes already issued, and succeeded so far that few copies of the first edition remain. His small fortune, which had been seriously impaired in bringing out his grand work, was exhausted in the effort to regain what was already issued, as the price rapidly advanced in proportion as fewer copies remained; and, mortified at the failure of his life's work, he died in poverty. It is said that some of his family, dissatisfied with the misfortune brought upon them by this disgrace and the loss of their patrimony, used a remaining copy for the production of a second edition, which met with a large sale, sufficient to repair the previous loss and restore the family fortune. This work of Beringer, in the end, exerted an excellent influence upon the dawning science of fossil remains. Observers became more cautious in announcing supposed discoveries, and careful study of natural objects gradually replaced vague hypotheses.

The above works, however, are hardly fair examples of the literature on fossils during this part of the eighteenth century. Scheuchzer had previously published his well-known "Complaint and Vindication of the Fishes," illustrated with good plates. Moro, in his work on "Marine Bodies which are found in the Mountains," 1740, showed the effects of volcanic action in elevating strata, and causing faults. Vallisneri had studied with care the marine deposits of Italy. Donati, in 1750, had investigated the Adriatic, and ascertained by soundings that shells and corals were being imbedded in the deposits there, just as they were found in the rocks.

John Gesner's dissertation, "De Petrificatis," published at Leyden in 1758, was a valuable contribution to the science. He enumerated the various kinds of fossils, and the different conditions in which they are found petrified, and stated that some of them, like those at Oeningen, resembled the shells, fishes, and plants of the neighboring region, while others, such as Ammonites and Belemnites, were either unknown species, or those found only in distant seas. He discusses the structure of the earth at length, and speculates as to the causes of changes in sea and land. He estimates that, at the observed rate of recession of the ocean, to allow the Apennines, whose summits are filled with marine shells, to reach their present height, would have taken about eighty thousand years, a period more than "ten times greater than the age of the universe." He accordingly refers the change to the direct command of the Deity, as related by Moses, that "the waters should be gathered together in one place, and the dry land appear."

Voltaire (1694-1778) discussed geological questions and the nature of fossils in several of his works, but his published opinions are far from consistent. He ridiculed effectively and justly the cosmogonists of his day, and showed also that he knew the true nature of organic remains. Finding, however, that theologians used these objects to confirm the Scriptural account of the deluge, he changed his views, and accounted for fossil shells found in the Alps by suggesting that they were Eastern species, dropped by the pilgrims on their return from the Holy Land!

Buffon, in 1749, published his important work on natural history, and included in it his "Theory of the Earth," in which he discussed, with much ability, many points in geology. Soon after the book was published, he received an official letter from the Faculty of Theology in Paris, stating that fourteen propositions in his works were reprehensible, and contrary to the creed of the Church. The first objectionable proposition was as follows: "The waters of the sea have produced the mountains and valleys of the land; the waters of the heavens, reducing all to a level, will at last deliver the whole land over to the sea; and the sea, successively prevailing over the land, will leave dry new continents like those we inhabit."

Buffon was politely requested by the college to recant, and, having no particular desire to be a martyr to science, submitted the following declaration, which he was required to publish in his next work: "I declare that I had no intention to contradict the text of Scripture; that I believe most firmly all therein related about the creation, both as to order of time and matter of fact; and I abandon everything in my book respecting the formation of the earth, and, generally, all which may be contrary to the narration of Moses."

This single instance will suffice to indicate one great obstacle to the advancement of science, even up to the middle of the eighteenth century.

Another important work appeared in France about this time, Bourguet's *"Traité des*

*Pétrifactions*," published in 1758, which is well illustrated with faithful plates. In England, a discourse on earthquakes, by Dr. Robert Hooke, was published in 1705. This author held some views in advance of his time, and maintained that figured stones were "really the several bodies they represent or the moldings of them petrified, and not, as some have imagined, a *lusus naturæ,* sporting herself in the needless formation of useless things." He anticipates one important conclusion from fossils, when he states that "though it must be very difficult to read them and to raise a chronology out of them, and to state the intervals of time wherein such or such catastrophes and mutations have happened, yet it is not impossible." He also states that fossil turtles, and such large Ammonites as are found in Portland, seem to have been the productions of hotter countries, and hence it is necessary to suppose that England once lay under the sea within the torrid zone. He seems to have suspected that some of the fossils of England belonged to extinct species, but thought they might possibly be found living in the bottom of distant oceans.

Dr. Woodward's "Natural History of the Fossils of England" appeared in 1729. This work was based on a systematic collection of fossils which he had brought together, and which he subsequently bequeathed to the University of Cambridge, where it is still preserved, with his arrangement carefully retained. The descriptive part of this work is interesting, but his conclusions are made to coincide strictly with the Scriptural account of the creation and deluge. He had previously stated, in another work, that he believed "the whole terrestrial globe to have been taken to pieces and dissolved at the flood, and the strata to have settled down from this promiscuous mass." In support of this view, he stated that "marine bodies are lodged in the strata according to the order of their gravity, the heavier shells in stones, the lighter in chalk, and so of the rest."

The most important work on fossils published in Germany at this time was that of George Wolfgang Knorr, which was continued after his death by Walch. This work consisted of four folio volumes, with many plates, and was printed at Nuremberg, 1755-'73. A large number of fossils were accurately figured and described, and the work is one of permanent value. A French translation of this work appeared in 1767-'78. Burton's "Oryctographie de Bruxelles," 1784, contains figures and descriptions of fossils found in Belgium.

Abraham Gottlieb Werner (1750-1817), Professor of Mineralogy at Freyberg, did much to advance the science of geology, and indirectly that of fossils. He first indicated the relations of the main formations to each other, and, according to his pupil, Professor Jameson, first made the highly important observation that "different formations can be discriminated by the petrifactions they contain." Moreover, that "the petrifactions contained in the oldest rocks are very different from any of the species of the present time; that the newer the formation, the more do the remains approach in form to the organic beings of the present creation." Unfortunately, Werner published little, and his doctrines were mainly disseminated by his enthusiastic pupils.

The great contest between the Vulcanists and the Neptunists started at this time, mainly through Werner, whose doctrines led to the controversy.

The comparative merits of fire and water, as agencies in the formation of certain rocks, were discussed with a heat and acrimony characteristic of the subject and the time. Werner believed in the aqueous theory, while the igneous theory was especially advocated by Hutton, of Edinburgh, and his illustrator, Playfair. This discussion resulted in the advancement of descriptive geology, but the study of fossils gained little thereby.

The "Protogæa" of Leibnitz, the great mathematician, published in 1749, about thirty years after his death, was a work of much merit. This author supposed that the earth had gradually cooled from a state of igneous fusion, and was subsequently covered with water. The subsidence of the lower part of the earth, the deposits of sedimentary strata from inundations, and their induration, as well as other changes, followed. All this he supposed to have been accomplished in a period of six natural days. In the same work Leibnitz shows that he had examined fossils with considerable care.

Linnæus (1707-1778), the famous Swedish botanist, and the founder of the modern system of nomenclature in natural history, confined his attention almost entirely to the living forms. Although he was familiar with the literature of fossil remains, and had collected them himself, he did not include them in his system of plants and animals, but kept them separate, with the minerals; hence he did little directly to advance this branch of science.

During the last quarter of the eighteenth century, the belief that fossil remains were deposited by the deluge sensibly declined, and the dawn of a new era gradually appeared. Let us pause for a moment here, and see what real progress had been made—what foundation had been laid on which to establish a science of fossil remains.

The true nature of these objects had now been clearly determined. They were the remains of animals and plants. Most of them certainly were not the relics of the Mosaic deluge, but had been deposited long before, part in fresh water and part in the sea. Some indicated a mild climate, and some the tropics. That any of these were extinct species was as yet only suspected. Large collections of fossils had now been made, and valuable catalogues, well illustrated, had been published. Something was known, too, of the geological position of fossils. Steno, long before, had observed that the lowest rocks were without life. Lehmann had shown that above these primitive rocks, and derived from them, were the secondary strata, full of the records of life; and above these were alluvial deposits, which he referred to local floods, and the deluge of Noah. Rouelle, Fuchsel, and Odoardi had shed new light on this subject. Werner had distinguished the transition rocks containing fossil remains, between the primitive and the secondary, while everything above the chalk he grouped together, as the "overflowed land." Werner, as we have seen, had done more than this, if we give him the credit his pupils claim for him. He had found that the formations he examined contained each its own peculiar fossils, and that from the older to the newer there was a gradual approach to recent forms. William Smith had worked out the same thing in England, and should equally divide the honor of this important discovery.

The greatest advance, however, up to this time, was that men now preferred to *observe* rather than to *believe,* and facts were held in greater esteem than vague speculations. With this preparation for future progress, the second period in the history of paleontology, as I have divided it, may appropriately be considered at an end.

Thus far, I have said nothing in regard to one branch of my subject, the *methods* of paleontological research, for, up to this time, of method there was none. We have seen that those of the ancients who noticed marine shells in the solid rock called them such, and concluded that they had been left there by the sea. The discovery of fossils led directly to theories of how the earth was formed. Here the progress was slow. Subterranean spirits were supposed to guard faithfully the mysteries of the earth; while above the earth. Authority guarded with still greater power the secrets men in advance of their age sought to know. The dominant idea of the first sixteen centuries of the present era was, that the universe was made for man. This was the great obstacle to the correct determination of the position of the earth in the universe, and, later, of the age of the earth. The contest of astronomy against authority was long and severe, but the victory was at last with science. The contest of geology against the same power followed, and continued almost to our day. The result is still the same. In the early stages of this contest there was no strife, for science was benumbed by the embrace of superstition and creed, and little could be done till that was cast off. In a superstitious age, when every natural event is referred to a supernatural cause, science cannot live; and often as the sacred fire may be kindled by courageous, far-seeing souls, will it be quenched by the dense mists of ignorance around it. Scarcely less fatal to the growth of science is the age of Authority, as the past proves too well. With freedom of thought came definite knowledge and certain progress; but two thousand years was long to wait.

With the opening of the present century began a new era in paleontology, which we may here distinguish as the third period in its history. This branch of knowledge became now a science. Method replaced disorder, and systematic study superseded casual observation. For the next half century the advance was continuous and rapid. One characteristic of this period was, the *accurate determination of fossils by comparison with living forms.* This will separate it from the two former epochs. Another distinctive feature of this period was the general belief that *every species, recent and extinct, teas a separate creation.*

At the very beginning of the epoch we are now to consider, three names stand out in bold relief—Cuvier, Lamarck, and William Smith. To these men the science of paleontology owes its origin. Cuvier and Lamarck, in France, had all the power which great talent, education, and station could give; William Smith, an English surveyor, was without culture or influence. The last years of the eighteenth century had been spent by each of these men in preparation for his chosen work, and the results were now given to the world. Cuvier laid the foundation of the paleontology of vertebrate animals; Lamarck, of the invertebrates; and Smith established the principles of stratigraphical paleontology. The investigator of fossils to-day seldom needs to consult earlier authors of the science.

George Cuvier (1769-1832), the most famous naturalist of his time, was led to the study of extinct animals by ascertaining that the remains of fossil elephants which he examined were extinct species. "This idea," he says later, "which I announced to the Institute in the month of January, 1796, opened to me views entirely new respecting the theory of the earth, and determined me to devote myself to the long researches and to the assiduous labors which have now occupied me for twenty-five years."

It is interesting to note here that in this first investigation of fossil vertebrates, Cuvier employed the same method that gave him such important results in his later researches.

Remains of elephants had been known to Europe for centuries, and many authors, from Pliny down to the contemporaries of Cuvier, had written about them. Some had regarded the bones as those of human giants, and those who recognized what they were considered them remains of the elephants imported by Hannibal or the Romans. Cuvier, however, compared the fossils directly with the bones of existing elephants, and proved them to be distinct. The fact that these remains belong to extinct species was of great importance. In the case of fossil shells, it was difficult to say that any particular form was not living in a distant ocean; but the two species of existing elephants, the Indian and the African, were well known, and there was hardly a possibility that another living one would be found.

It is important to bear in mind, too, that Cuvier's preparation for the study of the remains of animals was far in advance of any of his predecessors. He had devoted himself for years to careful dissections in the various classes of the animal kingdom, and was really the founder of comparative anatomy, as we now understand it. Cuvier investigated the different groups of the whole kingdom with care, and proposed a new classification, founded on the plan of structure, which in its main features is the one in use to-day. The first volume of his "Comparative Anatomy" appeared in 1800, and the work was completed in five volumes in 1805.

Previous to Cuvier, the only general catalogue of animals was contained in Linnæus's "Systema Naturæ." In this work, as we have seen, fossil remains were placed with the minerals, not in their appropriate places among the animals and plants. Cuvier enriched the animal kingdom by the introduction of fossil forms among the living, bringing all together into one comprehensive system. His great work, "Le Règne Animal," appeared in four volumes in 1817, and with its two subsequent editions remains the foundation of modern zoölogy. Cuvier's classic work on vertebrate fossils—"Recherches sur les Ossemens Fossiles," in four volumes, appeared in 1812-'13. Of this work it is but just to say that it could only have been written by a man of genius, profound knowledge, the greatest industry, and with the most favorable opportunities.

The introduction to this work was the famous "Discourse on the Revolutions of the Surface of the Globe," which has perhaps been as widely read as any other scientific essay.

The discovery of fossil bones in the gypsum-quarries of Paris by the workmen, who considered them human remains; the careful study of these relics by Cuvier, and his restorations from them of strange beasts that had lived long before, is a story with which you are all familiar. Cuvier was the first to prove that the earth had been inhabited by a succession of different series of animals, and he believed that those of each period were peculiar to the age in which they lived.

In looking over his work after a lapse of three quarters of a century, we can now see that Cuvier was wrong on some important points, and failed to realize the direction in which science was rapidly tending. With all his knowledge of the earth, he could not free himself from tradition, and believed in the universality and power of the Mosaic deluge. Again, he refused to admit the evidence brought forward by his distinguished colleagues against the permanence of species, and used all his great influence to crush out the doctrine of evolution, then first proposed. Cuvier's definition of a species, the dominant one for half a century, was as follows: "A species comprehends all the individuals which descend from each other, or from a common parentage, and those which resemble them as much as they do each other."

The law of "Correlation of Structures," as laid down by Cuvier, has been more widely accepted than almost anything else that bears his name; and yet, although founded in truth, and useful within certain limits, it would certainly lead to serious error if applied widely in the way he proposed.

In his discourse he sums up this law as follows: "A claw, a shoulder-blade, a condyle, a leg or arm bone, or any other bone separately considered, enables us to discover the description of teeth to which they have belonged; so also reciprocally we may determine the form of the other bones from the teeth. Thus, commencing our investigation by a careful survey of any one bone by itself, a person who is sufficiently master of the laws of organic structure may, as it were, reconstruct the whole animal to which that bone had belonged."

We know to-day that unknown extinct animals cannot be restored from a single tooth or claw, unless they are very similar to forms already known. Had Cuvier himself applied his methods to many forms from the early Tertiary or older formations, he would have failed. If, for instance, he had had before him the disconnected fragments of an Eocene Tillodont, he would undoubtedly have referred a molar tooth to one of his pachyderms; an incisor tooth to a rodent; and a claw-bone to a carnivore. The tooth of a Hesperornis would have given him no possible hint of the rest of the skeleton, nor its swimming feet the slightest clew to the ostrich-like sternum or skull. And yet the earnest belief in his own methods led Cuvier to some of his most important discoveries.

Jean Lamarck (1744-1829), the philosopher and naturalist, a colleague of Cuvier, was a learned botanist before he became a zoologist. His researches on the invertebrate fossils of the Paris Basin, although less striking, were not less important than those of Cuvier on the vertebrates; while the conclusions he derived from them form the basis of modern biology. Lamarck's method of investigation was the same essentially as that used by Cuvier, namely, a direct comparison of fossils with living forms. In this way he soon ascertained that the fossil shells imbedded in the strata beneath Paris were many of them extinct species, and those of different strata differed from each other. His first memoir on this subject appeared in 1802, and, with his later works, effected a revolution in conchology. His "System of Invertebrate Animals" appeared the year before, and his famous "Philosophie Zoölogique" in 1809. In these two works, Lamarck first announced the principles of evolution. In the first volume of his "Natural History of Invertebrate Animals" he gave his theory in detail; and today one can only read with astonishment his far-reaching anticipations of modern science. These views were strongly supported by Geoffroy Saint-Hilaire, but bitterly opposed by Cuvier; and their great contest on this subject is well known.

In looking back from this point of view, the philosophical breadth of Lamarck's conclusions, in comparison with those of Cuvier, is clearly evident. The invertebrates on which Lamarck worked offered less striking evidence of change than the various animals investigated by Cuvier; yet they led Lamarck directly to evolution, while Cuvier ignored what was before him on this point, and rejected the proof offered by others. Both pursued the same methods, and had an abundance of material on which to work, yet the facts observed induced Cuvier to believe in catastrophes, and Lamarck in the uniform course of nature. Cuvier declared species to be permanent, Lamarck that they were descended from others. Both men stand in the first rank in science, but Lamarck was the prophetic genius half a century in advance of his time.

II

While the Paris Basin was yielding such important results for paleontology, its geological structure was being worked out with great care. The results appeared in a volume by Cuvier and Alexandre Brongniart, chiefly the work of the latter, published in 1808. This was the first systematic investigation of Tertiary strata. Three years later, the work was issued in a more extended form. The separate formations were here carefully distinguished by their fossils, the true importance of which for this purpose being distinctly recognized. This advance was not accepted without some opposition, and it is an interesting fact that Jameson, who claimed for Werner the theory here put in practice, rejected its application, and wrote as follows: "To Cuvier and Brongniart we are indebted for much valuable information in their description of the country around Paris, but we must protest against the use they have made of fossil organic remains in their geognostical descriptions and investigations."

William Smith (1769-1839), "the father of English geology," had previously published a

"Tabular View of the British Strata." He appears to have arrived independently at essentially the same view as Werner in regard to the relative position of stratified rocks. He had determined that the order of succession was constant, and that the different formations might be identified at distant points by the fossils they contained. In his later works, "Strata identified by Organized Fossils," published in 1816-'20, and "Stratigraphical System of Organized Fossils," 1817, he gave to the world results of many years of careful investigations on the Secondary formations of England. In the latter work, he speaks of the success of his method in determining strata by their fossils, as follows: "My original method of tracing the strata by the organized fossils imbedded therein is thus reduced to a science not difficult to learn. Ever since the first written account of this discovery was circulated in 1799, it has been closely investigated by my scientific acquaintances in the vicinity of Bath, some of whom search the quarries of different strata in that district with as much certainty of finding the characteristic fossils of the respective rocks as if they were on the shelves of their cabinets."

The systematic study of fossils now attracted attention in England also, and was prosecuted with considerable zeal, although with less important results than in France. An extensive work on this subject, by James Parkinson, entitled "Organic Remains of a Former World," was begun in 1804, and completed in three volumes in 1811. A second edition appeared in 1833. This work was far in advance of previous publications in England, and, being well illustrated, did much to make the collection and study of fossils popular. The belief in the geological effects of the deluge had not yet lost its power, although restricted now to the later deposits; for Parkinson, in his later edition, wrote as follows: "Why the earth was at first so constituted that the deluge should be rendered necessary—why the earth could not have been at first stored with all those substances and endued with all those properties which seemed to have proceeded from the deluge—why so many beings were created, as it appears, for the purpose of being destroyed—are questions which I presume not to answer."

William Buckland (1784-1856) published in 1823 his celebrated "Reliquiæ Diluvianæ," in which he gave the results of his own observations in regard to the animal remains found in the caves, fissures, and alluvial gravels of England. The facts presented are of great value, and the work was long a model for similar researches. Buckland's conclusions were, that none of the human remains discovered in the caves were as old as the extinct mammals found with them, and that the deluge was universal. In speaking of fossil bones found in the Himalaya Mountains, he says: "The occurrence of these bones at such an enormous elevation in the region of eternal snow, and consequently in a spot now unfrequented by such animals as the horse and deer, can, I think, be explained only by supposing them to be of antediluvian origin, and that the carcasses of the animals were drifted to their present place, and lodged in sand, by the diluvial waters."

The foundation of the "Geological Society of London," in 1807, marks an important point in the history of paleontology. To carefully collect materials for future generalizations was the object in view, and this organization gradually became the center in Great Britain for those interested in geological science. The society was incorporated in 1826, and has since been the leading organization in Europe for the advancement of the sciences within its field. The Geological Society of France, established at Paris in 1832, and the German Geological Society, founded at Berlin in 1848, have likewise contributed largely to geological investigations in these countries, and to some extent in other parts of the world. In the publications of these three societies the student of paleontology will find a mine of valuable materials for his work.

The systematic study of fossil plants may be said to date from the publication of Adolphe Brongniart's "Prodrome," in 1828. This was very soon followed by his larger work, "Histoire des Végétaux Fossiles," issued in 1828-'48. Brongniart pursued the same method as Cuvier and Lamarck, viz., the comparison of fossils with living forms, and his results were of great importance. In his "Tableau des Genres Végétaux Fossiles," etc., published in Paris in 1849, he gives the classification and distribution of the genera of fossil plants, and traces out the historical progression of vegetable life on the globe, as he had done to a great extent in his previous works. He shows that the cryptogamic forms prevailed in the Primary formations, the conifers and cycads in the Secondary, and the higher forms in the Tertiary, while four fifths of living plants are exogens.

In England, Lindley and Hutton published, in 1831-'37, a valuable work in three volumes, entitled "Fossil Flora of Great Britain." This work was illustrated by many accurate plates, in which the plants of the coal formation were especially represented, Henry Witham also published two works in 1831 and 1833, in which he treated especially of the internal structure of fossil plants. "Antediluvian Phytology," by Artis, was published in London in 1838, Bowerbank's "History of the Fossil Fruits and Seeds of the London Clay" appeared in 1843, Hooker's memoir "On the Vegetation of the Carboniferous Period as compared with that of the Present Day," published in 1848, was an important contribution to the science. Bunbury, Williamson, and others, also published various papers on fossil plants. This branch of paleontology, however, attracted much less attention in England than on the Continent.

In Germany, the study of fossil plants dates back to the beginning of the century. Von Schlotheim, a pupil of Werner, published in 1804 an illustrated volume on this subject. A more important work was that of Count Sternberg, issued in 1820-'38, and illustrated with excellent plates. Cotta, in 1832, published a book with the title "Die Dendrolithen," in which he gave the results of his investigations on the inner structure of fossil plants. Von Gutbier, in 1835, and Germar, in 1844-'53, described and figured the plants of two important localities in Germany. Corda's "Beiträge zur Flora der Vorwelt," issued at Prague, in 1845, was essentially a continuation of the work of Stemberg, Unger's "Chloris Protogæa," 1841-45, "Genera et Species Plantarum Fossilium," 1850, and his larger work, published in 1852, are all standard authorities. It the latter, the theory of descent is applied to the vegetable world. Schimper and Mougeot's "Monograph on the Fossil Plants of the Vosges," 1845, was well illustrated, and contained noteworthy results.

Göppert in 1836 published a valuable memoir entitled "Systema Filicum Fossilium," in which he made known the results of his study of fossil ferns. In the same year, this botanist began a series of experiments in which he attempted to imitate the process of fossilization, as found in nature. He steeped various animal and vegetable substances in waters holding, some calcareous, others siliceous, and others metallic matter in solution. After a slow saturation, the substances were dried, and exposed to heat until the organic matters were burned. In this way Göppert successfully imitated various processes of petrifaction, and explained many things in regard to fossils that had previously been in question. His discovery of the remains of plants throughout the interior of coal did much to clear up the doubts about the formation of that substance. In 1841 Göppert published an important work in which he compared the genera of fossil plants with those now living. In 1852 another extensive work by this author appeared, entitled "Fossile Flora des Uebergangs-Gebirges."

Andræ, Braun, Dunker, Ettingshausen, Geinitz, and Goldenberg, all made notable contributions to fossil botany in Germany during the period we are now considering.

The systematic study of invertebrate fossils, so admirably begin by Lamarck, was continued actively in France. The Tertiary shells of the Seine Valley were further investigated by Defrance, and especially by Deshayes, whose great work on this subject was begun in 1824. Des Moulins's essay on "Sphérulites" in 1826, Blainville's memoir on "Belemnites" in 1827, Ferussac's various memoirs on land and freshwater fossil shells, were valuable additions to the subject. A later work of great importance was D'Orbigny's "Paléontologie Française," 1840-'44, which described the mollusca and radiates in detail, according to formations. The other publications of this author are both numerous and valuable. Brongniart and Desmarest's "Histoire naturelle des Crustacés Fossiles," published in 1822, is a pioneer work on this subject. Michelin's memoir on the fossil corals of France, 1841-'46, was another important contribution to paleontology. Agassiz's works on fossil Echinoderms and Mollusks are valuable contributions to the science. The works of D'Archiac, Coquand, Cotteau, Desor, Edwards, Haime, and De Verneuil, are likewise of permanent value.

In Italy, Bellardi, Merian, Michellotti, Phillipi, Zigno, and others, contributed important results to paleontology.

In Belgium, Bosquet, Nyst, Koninck, Ryckholt, Van Beneden, and others have all aided materially in the progress of the science.

In England, also, invertebrate fossils were studied with care, and continued progress was made. Sowerby's "Mineral Conchology of Great Britain," in six volumes, a systematic work of great value, was published in 1812-'30, and soon after was translated into French and German. Its figures of fossil shells are excellent, and it is still a standard work. Miller's "Natural History of the Crinoidea," published at Bristol, in 1821, and Austin's later monograph, are valuable for reference. Brown's "Fossil Conchology of Britain and Ireland" appeared in 1839, and Brodie's "History of the Fossil Insects of England" in 1845. Phillips's illustration of the geology of Yorkshire, 1829-'36, and his work on the Palæozoic fossils of Cornwall, Devonshire, and West Somerset, 1843, contained a great deal of original matter in regard to fossil remains. Morris's "Catalogue of British Fossils," issued in 1843, and the later edition in 1854, is most useful to the working paleontologist. The memoirs of Davidson on the Brachiopoda, Edwards, Forbes, Morris, Lycett, Sharpe, and Wood on other Mollusca, Wright on the Echinoderms, Salter on Crustacea, Busk on Polyzoa, Jones on the Entomostraca, and Duncan and Lonsdale on Corals, are of especial value. King's volume on Permian fossils, Mantell's various memoirs, Dixon's work on the fossils of Sussex, 1850, and McCoy's works on Palæozoic fossils, all deserve honorable mention. Sedgwick, Murchison, and Lyell, although their greatest services were in systematic geology, each contributed important results to the kindred science of paleontology during the period we are reviewing.

In Germany, Schlotheim's treatise, "Die Petrifactenkunde," published at Gotha in 1820, did much to promote a general interest in fossils. By far the most important work issued on this subject was the "Petrifacta Germanica," by Goldfuss, in three folio volumes, 1826-1844, which has lost little of its value. Bronn's "Geschichte der Natur," 1841-'46, was a work of great labor, and one of the most useful in the literature of this period. The author gave a list of all the known fossil species, with full references, and also their distribution through the various formations. This gave exact data on which to base generalizations, hitherto of comparatively little value.

Among other early works of interest in this department may be mentioned Dalman's memoir on "Trilobites," 1828, and Burmeister's on the same subject, 1843. Giebel's well-known "Fauna der Vorwelt," 1847-'56, gave lists of all the fossils described up to that time, and hence is a very useful work. The "Lethæa Geognostica," by Bronn, 1834-'38, and the second edition by Bronn and Roemer, 1846-'56, is a comprehensive general treatise on paleontology, and the most valuable work of the kind yet published.

The researches of Ehrenberg, in regard to the lowest forms of animals and plants, threw much light on various points in paleontology, and showed the origin of extensive deposits, the nature of which had before been in doubt. Von Buch, Barrando, Beyrich, Berendt, Dunker, Geinitz, Heer, Homes, Klipstein, Von Münster, Reuss, Roemer,

Sandberger, Suess, Von Hagenow, Von Hauer, Zeiten, and many others, all aided in the advancement of this branch of science. Angelin, Hisinger, and Nilsson, in Scandinavia; Abich, de Waldheim, Eichwald, Keyserling, Kutorga, Nordmann, Pander, Rouillier, and Volborth, in Russia; and Pusch in Poland, published important results on fossil invertebrates.

The impetus given by Cuvier to the study of vertebrate fossils extended over Europe, and great efforts were made to continue discoveries in the direction he had so admirably pointed out. *Louis Agassiz* (1807-1873), a pupil of Cuvier, and long an honored member of this association, attained eminence in the study of ancient as well as of recent life. His great work on Fossil Fishes deserves to rank next to Cuvier's "Ossemens Fossiles." The latter contained mainly fossil mammals and reptiles, while the fishes were left without an historian till Agassiz began his investigations. His studies had admirably fitted him for the task, and his industry brought together a vast array of facts bearing on the subject. The value of this grand work consists not only in its faithful descriptions and plates, but also in the more profound results it contained. Agassiz first showed that there is a correspondence between the succession of fishes in the rocks and their embryonal development. This is now thought to be one of the strongest points in favor of evolution, although its discoverer interpreted the facts as bearing the other way.

Pander's memoirs on the fossil fishes of Russia form a worthy supplement to Agassiz's classic work. Brandt's publications are likewise of great value; and those of Lund, in Sweden, have an especial interest to Americans, in consequence of his researches in the caves of Brazil.

Croizet and Jobert's "Recherches sur les Ossemens Fossiles du Département du Puy-de-Dôme," published in 1828, contained valuable results in regard to fossil mammals. Geoffrey Saint-Hilaire's researches on fossil reptiles, published in 1831, were an important advance. De Serres and De Christol's explorations in the caverns in the south of France, published between 1829 and 1839, were of much value. Schmerling's researches in the caverns of Belgium, published in 1833-'36, were especially important on account of the discovery of human remains mingled with those of extinct animals. Deslongchamp's memoirs on fossil reptiles, 1835, are still of great interest. Pictet's general treatise on paleontology was a valuable addition to the literature, and has done much to encourage the study of fossils. De Blainville, in his grand work, "Ostéographie," issued in 1839-'56, brought together the remains of living and extinct vertebrates, forming a series of the greatest value for study. Aymard and Pomel's contributions to vertebrate paleontology are both of value. Gervais and Lartet added much to our knowledge of the subject, and Bravard and Hebert's memoirs are well known.

The brilliant discoveries of Cuvier in the Paris Basin excited great interest in England, and, when it was found that the same Tertiary strata existed in the south of England, careful search was made for vertebrate fossils. Remains of some of the same genera described by Cuvier were soon discovered, and other extinct animals new to science were found in various parts of the kingdom. König, to whom we owe the name *Ichthyosaurus,* and Conybeare, who gave the generic designation

*Plesiosaurus,* and also *Mosasaurus,* were among the earliest writers in England on fossil reptiles. The discovery of these three extinct types, and the discussion as to their nature form a most interesting chapter in the annals of paleontology. The discovery of the *Iguanodon,* by Mantell, and the *Megalosaurus,* by Buckland, excited still higher interest. These great reptiles differed much more widely from living forms than the mammals described by Cuvier, and the period in which they lived soon became known as the "age of Reptiles." The subsequent researches of these authors added largely to the existing knowledge of various extinct forms, and their writings did much to arouse public interest in the subject.

Richard Owen, a pupil of Cuvier, followed, and brought to bear upon the subject an extensive knowledge of comparative anatomy, and a wide acquaintance with existing forms. His contributions have enriched almost every department of paleontology, and of extinct vertebrates especially, he has been, since Cuvier, the chief historian. The fossil reptiles of England he has systematically described, as well as those of South Africa. The extinct Struthious birds of New Zealand he has made known to science, and accurately described in extended memoirs. His researches on the fossil mammals of Great Britain, the extinct Edentates of South America, and the ancient Marsupials of Australia, each forms an important chapter in the history of our science.

The personal researches of Falconer and Cautley in the Siwalik Hills of India brought to light a marvelous vertebrate fauna of Pliocene age. The remains thus secured were made known in their great work, "Fauna Antiqua Sivalensis," published at London in 1845. The important contributions of Egerton to our knowledge of fossil fishes, and Jardine's well-known work, "Ichnology of Annandale," also belong to this period.

The study of vertebrate fossils in Germany was prosecuted with much success during the present period. Blumenbach, the ethnologist, in several publications between 1803 and 1814, recorded valuable observations on this subject. In 1812 Sömmering gave an excellent figure of a pterodactyl, which he named and described. Goldfuss's researches on the fossil vertebrates from the caves of Germany, published in 1820-'23, made known the more important facts of that interesting fauna. His later publications on extinct amphibians and reptiles were also noteworthy. Jäger's investigations on the extinct vertebrate fauna of Würtemberg, published between 1824 and 1839, were an important advance. To Plieninger's researches in the same regiori, 1834-'44, we owe the discovery of the first Triassic mammal (*Microlestes*), as well as important information in regard to Labyrinthodonts. Kaup's researches on fossil mammals, 1832-'41, brought to light many interesting forms, and to him we are indebted for the generic name *Dinotherium,* and excellent descriptions of the remains then known.

Count Münster's "Beiträge zur Petrifactenkunde," published 1843-'46, contained several valuable papers on fossil vertebrates; and the separate papers by the same author are of interest. Andreas "Wagner wrote on Pterosaurians in 1837, and later gave the first description of fossil mammals of the Tertiary of Greece, 1837-'40. Johannes Müller published an important illustrated work on the Zeuglodonts in 1849, and various notable memoirs; and Quenstedt interesting descriptions of fossil reptiles, as well as other papers of much value. Rütimeyer's suggestive memoirs are widely known.

Hermann von Meyer's contributions to vertebrate paleontology are by far the most important published in Germany during the period we are now considering. From 1830 his investigations on this subject were continuous for nearly forty years, and his various publications are all of value. His "Beiträge zur Petrifactenkunde," 1831-'33, contains a series of valuable memoirs. His "Palæologica," issued in 1832, includes a synopsis of the fossil vertebrates then known, with much original matter. His great work, "Zur Fauna der Vorwelt," 1845-'60, includes a series of monographs invaluable to the student of vertebrate paleontology. This work, as well as his other chief publications, was illustrated with admirable plates from his own drawings. Other memoirs by this author will be found in the "Palæontographica," of which he was one of the editors. In the many volumes of this publication, which began in 1851, and is still continued, will be found much to interest the investigator in any branch of paleontology.

The "Palæntographical Society of London," established in 1847, has also issued a series of volumes containing valuable memoirs in various branches of paleontology. These two publications together are a storehouse of knowledge in regard to extinct forms of animal and vegetable life.

It may be interesting here to note briefly the use of general terms in paleontology, as the gradual progress of the science was indicated to some extent in its terminology. At first, and for a long time, the name "*fossil*" was appropriately used for objects dug from the earth, both minerals and organic remains. The term "oryctology," having essentially the same meaning, was also used for this branch of study. For a long period, too, the termination *ites* (*λίθος*, a stone) was applied to fossils to distinguish them from the corresponding living forms; as, for instance, "ostracites", used by Pliny. At a later date the general name "figured stones" (*lapides figurati*) was extensively used; and, less frequently, "deluge-stones" (*lapides diluviani*). The term "organized fossils" was used to distinguish fossils from minerals when the real difference became known, although the name "*reliquiæ*" was sometimes employed. The term "petrifactions" (*petrificata*) was defined by John Gesner in his work on fossils in 1758, and was afterward extensively used. Paleontology is comparatively a modern term, having only come into use only within the last half century. It was introduced about 1830, and soon was generally adopted in France and England; but in Germany it met with less favor, though used to some extent.

It would be interesting, too, did time permit, to trace the various opinions and superstitions held at different times in regard to some of the more common fossils, for example the Ammonite or the Belemnite—of their supposed celestial origin; of their use as medicine by the ancients, and in the East to-day; of their marvelous power as charms among the Romans, and still among the American Indians. It would be instructive, also, to compare the various views expressed by students in science concerning some of the stranger extinct forms—for instance, the Nummulites, among Protozoa; the Rudistes, among Mollusks; or the Mosasaurus, among reptiles. Dissimilar as such views were, they indicate in many cases gropings after truth—natural steps in the increase of knowledge.

The third period in the history of paleontology, which, as I have said, began with Cuvier and Lamarck at the beginning of the present century, forms a natural epoch extending through six decades. The definite characteristics of this period, as stated, were dominant during all this time, and the progress of paleontology was commensurate with that of intelligence and culture.

For the first half of this period, the marvelous discoveries in the Paris Basin excited astonishment and absorbed attention; but the real significance and value of the facts made known by Cuvier, Lamarck, and William Smith were not appreciated. There was still a strong tendency to regard fossils merely as interesting objects of natural history, as in the previous period, and not as the key to profounder problems in the earth's history. Many prominent geologists were still endeavoring to identify formations in different countries by their mineral characters rather than by the fossils imbedded in them. Such names as "Old Red Sandstone" and

"New Red Sandstone" were given in accordance with this opinion. Humboldt, for example, attempted to compare the formations of South America and Europe by their mineral features, and doubted the value of fossils for this purpose. In 1823 he wrote as follows: "Are we justified in concluding that all formations are characterized by particular species? that the fossil shells of the chalk, the muschelkalk, the Jura limestone, and the Alpine limestones are all different? I think this would be pushing the induction much too far." Jameson still thought minerals more important than fossils for characterizing formations; while Bakewell, later yet, defines paleontology as comprising "fossil zoölogy and fossil botany, a knowledge of which may appear to the student as having little connection with geology."

During the later half of the third period, greater progress was made, and before its close geology was thoroughly established as a science. Let us consider for a moment what had really been accomplished up to this time.

It had now been proved beyond question that portions at least of the earth's surface had been covered many times by the sea, with alternations of fresh water and of land; that the strata thus deposited were formed in succession, the lowest of the series being the oldest; that a distinct succession of animals and plants had inhabited the earth during the different geological periods; and that the order of succession found in one part of the earth was essentially the same in all. More than 30,000 new species of extinct animals and plants had now been described It had been found, too, that from the oldest formations to the most recent, there had been an advance in the grade of life, both animal and vegetable, the oldest forms being among the simplest, and the higher forms successively making their appearance.

It had now become clearly evident, moreover, that the fossils from the older formations were all extinct species, and that only in the most recent deposits were there remains of forms still living. The equally important fact had been established that in several groups of both animals and plants the extinct forms were vastly more numerous than the living, while several orders of fossil animals had no representatives in modern times. Human remains had been found mingled with those of extinct animals, but the association was regarded as an accidental one by the authorities in science; and the very recent appearance of man on the earth was not seriously questioned. Another important conclusion reached, mainly through the labors of Lyell, was, that the earth had not been subjected in the past to sudden and violent revolutions; but the great changes wrought had been gradual, differing in no essential respect from those still in progress. Strangely enough, the corollary to this proposition, that life, too, had been continuous on the earth, formed at that date no part of the common stock of knowledge.

In the physical world the great law of "correlation of forces" had been announced and widely accepted; but, in the organic world, the dogma of the miraculous creation of each separate species still held sway, almost as completely as when Linnæus declared, "There are as many different species as there were different forms created in the beginning by the Infinite Being." But the dawn of a new era was already breaking, and the third period of paleontology we may consider now at an end.

Just twenty years ago, science had reached a point when the belief in "special creations" was undermined by well-established facts, slowly accumulated. The time was ripe. Many naturalists were working at the problem, convinced that evolution was the key to the present and the past. But how had Nature brought this change about? While others pondered, Darwin spoke the magic words—"*natural selection*," and a new epoch in science began.

The fourth period in the history of paleontology dates from this time, and is the period of to-day. One of the main characteristics of this epoch is the belief that *all life, living and extinct, has been evolved from simple forms.* Another prominent feature is the accepted fact of the *great antiquity of the human race.* These are quite sufficient to distinguish this period sharply from those that preceded it.

The publication of Charles Darwin's work on the "Origin of Species," November, 1859, at once aroused attention, and started a revolution which has already in the short space of two decades changed the whole course of scientific thought. The theory of "natural selection," or, as Spencer has happily termed it, the "survival of the fittest," had been worked out independently by Wallace, who justly shares the honor of the discovery. We have seen that the theory of evolution was proposed and advocated by Lamarck, but he was before his time. The anonymous author of the "Vestiges of Creation," which appeared in 1844, advocated a somewhat similar theory, which attracted much attention, but the belief that species were immutable was not sensibly affected until Darwin's work appeared.

The difference between Lamarck and Darwin is essentially this: Lamarck proposed the theory of evolution; Darwin changed this into a doctrine, which is now guiding the investigations in all departments of biology. Lamarck failed to realize the importance of time, and the interaction of life on life. Darwin, by combining these influences with those also suggested by Lamarck, has shown *how* the existing forms on the earth may have been derived from those of the past.

This revolution has influenced paleontology as extensively as any other department of science, and hence the new period we are discussing. In the last epoch, species were represented independently, by parallel lines; in the present period, they are indicated by dependent, branching lines. The former was the analytic, the latter is the synthetic period. To-day, the animals and plants now living are believed to be genetically connected with those of the distant past; and the paleontologist no longer deems species of the first importance, but seeks for relationships and genealogies, connecting the past with the present. Working in this spirit, and with such a method, the advance during the last decade has been great, and is an earnest of what is yet to come.

The progress of paleontology in Great Britain during the present period has been great, and the general interest in the science much extended. The views of Darwin soon found acceptance here. Next to his discovery of "natural selection," Darwin was fortunate in having so able and bold an expounder as Huxley, who was one of the first to adopt his theory and give it a vigorous support. Huxley's masterly researches have been of great benefit to all departments of biology, and his contributions to paleontology are invaluable. Among the latter, his original investigations on the relations of birds and reptiles are especially noteworthy. His various memoirs on extinct reptiles, amphibians, and fishes belong to the permanent literature of the subject. The important researches of Owen on the fossil vertebrates have been continued to the present time. He has added largely to his previous publications on the British fossil reptiles, birds, and mammals, the extinct reptiles of South Africa, and the post-Tertiary birds of New Zealand. His description of the *Archæopteryx,* near the beginning of the period, was a most welcome contribution.

The investigations of Egerton on fossil fishes have likewise been continued with important results. Busk, Dawkins, Flower, and Sanford have made valuable contributions to the history of fossil mammals. Bell, Günther, Hulke, Lankester, Newton, Powrie, Miall, Tracquair, and Seely have made notable additions to our knowledge of reptiles, amphibians, and fishes. Among invertebrates, the Crustacea have been especially studied by Jones, Salter, and Woodward. Davidson, Etheridge, Lycett, Morris, Phillips, Wood, and Wright have continued their researches on the mollusks; Duncan, Nicholson, and others have investigated the extinct corals; and Binney, Carruthers, and Williamson the fossil plants. Numerous other important contributions have been made to the science in Great Britain during the present period.

On the Continent the advance in paleontology has, during the last two decades, been equally great. In France, Gervais continued his memoirs on extinct vertebrates nearly to the present date; while Gaudry has published several volumes on the subject that are models for all students of the science. His work on the fossil animals of Greece is a perfect monograph of its kind, and his later publications are all of importance. Lartet's various works are of permanent value, and his application of paleontology to archaeology brought notable results. The volume of Alphonse Milne-Edwards on fossil Crustacea was a fit supplement to Brongniart and Desmarest's well-known work; while his grand memoir on fossil birds deserves to rank with the classic volumes of Cuvier. Duvernoy, Filhol, Hébert, Sauvage, and others have also published interesting results on fossil vertebrates.

Van Beneden's researches on the fossil vertebrates of Belgium have produced results of great value. Pictet, Rütimeyer, and Wiedersheim in Switzerland; Bianconi, Carnalia, Forsyth-Major, and Sismonda in Italy; and Nodot in Spain, have likewise published important memoirs. The extinct vertebrates have been studied in Germany by Von Meyer, Cams, Fraas, Giebel, Heckel, Haase, Hensel, Kayser, Kner, Ludwig, Peters, Portis, Maack, Salenka, Zittel, and many others; in Holland by Winkler; in Denmark by Reinhardt; and in Russia by Brandt and Kowalewsky.

The fossil invertebrates have been investigated with care by D'Archiac, D'Orbigny, Bayle, Fromentel, Oustalet, and others in France; Desor, Loriol, Mayer, Ooster, and Roux in Switzerland; Capellini, Massalongo, Michellotti, Meneghini, and Sismonda in Italy; Barrande, Benecke, Beyrich, Dames, Dorn, Ehlers, Geinitz, Giebel, Gümbel, Feistmantel, Hagen, Von Hauer, Von Heyden, Von Fritsch, Laube, Oppel, Quenstedt, Roemer,

Schlüter, Suess, Speyer, and Zittel in Germany. The fossil plants have been studied in these countries by Massalongo, Saporta, Zigno, Fiedler, Goldenberg, Gehler, Heer, Goppert, Ludwig, Schimper, Schenk, and many others.

Among the recent researches in paleontology in other regions may be mentioned those of Blanford, Feistmantel, Lydekker, and Stoliczka in India; Haast and Hector in New Zealand; and Krefft and McCoy in Australia—all of whom have published valuable results.

Of the progress of paleontology in America I have thus far said nothing, and I need now say but little, as many of you are doubtless familiar with its main features. During the first and second periods in the history of paleontology, as I have defined them, America, for most excellent reasons, took no part. In the present century, during the third period, appear the names of Bigsby, Green, Morton, Mitchell, Rafinesque, Say, and Troost, all of whom deserve mention. More recently the researches of Conrad, Dana, Deane, DeKay, Emmons, Gibbes, Hitchcock, Holmes, Lea, McChesney, Owen, Redfield, Rogers, Rominger, Shumard, Swallow, and many others have enlarged our knowledge of the fossils of this country.

The contributions of James Hall to the invertebrate paleontology of this country form the basis of our present knowledge of the subject. The extensive labors of Meek in the same department are likewise entitled to great credit, and will form an important chapter in the history of the science. The memoirs of Billings, Gabb, Scudder, White, and Whitfield are numerous and important; and the publications of Derby, Hartt, Hyatt, James, Miller, Shaler, Rathbun, Vogdes, Whiteaves, and Winchell, are also of value. To Dawson, Lesquereux, and Newberry we mainly owe our present knowledge of the fossil plants of this country.

The foundation of our vertebrate paleontology was laid by Leidy, whose contributions have enriched nearly every department of the subject. The numerous publications of Cope are well known. Agassiz, Allen, Baird, Dawson, Deane, DeKay, Emmons, Gibbes, Harlan, Hitchcock, Jefferson, Lea, Le Conte, Newberry, Redfield, St. John, Warren, Whitney, Worthen, Wyman, and others have all added to our knowledge of

American fossil vertebrates. The chief results in this department of our subject I have already laid before you on a previous occasion, and hence need not dwell upon them here.

In this rapid sketch of the history of paleontology I have thought it best to speak of the earlier periods more in detail, as they are less generally known, and especially as they indicate the growth of the science, and the obstacles it had to surmount. With the present work in paleontology, moreover, you are all more or less familiar, as the results are now part of the current literature. To assign every important discovery to its author would have led me far beyond my present plan. I have only endeavored to indicate the growth of the science by citing the more prominent works that mark its progress, or illustrate the prevailing opinions and state of knowledge at the time they were written.

In considering what has been accomplished, directly or indirectly, it is well to bear in mind that without paleontology there would have been no science of geology. The latter science originated from the study of fossils, and not the reverse, as generally supposed. Paleontology, therefore, is not a mere branch of geology, but the foundation on which that science mainly rests. This fact is a sufficient excuse, if one were wanting, for noting the early opinions in regard to the changes of the earth's surface, as these changes were first studied to explain the position of fossils. The investigation of the latter first led to theories of the earth's formation, and thus to geology. When speculation replaced observation, fossils were discarded, and for a time the mineral characters of strata were thought to be the key to their position and age. For some time after this, geologists, as we have seen, apologized for using fossils to determine formations, but for the last half century their value for this purpose has been fully recognized.

The services which paleontology has rendered to botany and zoölogy are less easy to estimate, but are very extensive. The classification of these sciences has been rendered much more complete by the intercalation of many intermediate forms. The probable origin of various living species has been indicated by the genealogies suggested by extinct types; while our knowledge of the geographical distribution of animals and plants at the present day has been greatly improved by the facts brought out in regard to the former distribution of life on the globe.

Among the vast number of new species which have been added are the representatives of a number of new orders entirely unknown among living forms. The distribution of these extinct orders among the different classes is interesting, as they are mainly confined to the higher groups. Among the fossil plants no new orders have yet been found. There are none known among the Protozoa or the Mollusca. The Radiates have been enriched by the extinct orders of Blastoidea, Cystidea, and Edrioasterida; and the Crustaceans by the Eurypterida and Trilobita. Among the vertebrates, no extinct order of fossil fishes has yet been found; but the amphibians have been enlarged by the important order Labyrinthodonta, The greatest additions have been among the reptiles, where the majority of the orders are extinct. Here we have at the present date the Ichthyosauria, Sauranocontia, Plesiosauria, and Mosasauria, among the marine forms; the Pterosauria, including the Pteranodontia, containing the flying forms; and the Dinosauria, including the Sauropoda—the giants among reptiles; likewise the Dicynodontia, and probably the Theriodontia, among the terrestrial forms. Although but few fossil birds have been found below the Tertiary, we have already among the Mesozoic forms three new orders: the Saururæ, represented by *Archæopteryx;* the Odontotormæ, with *Ichthyornis* as the type; and the Odontoleæ, based upon *Hesperornis;* all of these orders being included in the sub-class Odontornithes, or toothed birds. Among mammals, the new groups regarded as orders are the Toxodontia and the Dinocerata, among the Ungulates; and the Tillodontia, including strange Eocene mammals whose exact affinities are yet to be determined.

Among the important results in vertebrate paleontology, are the genealogies, made out with considerable probability, for various existing animals. Many of the larger mammals have been traced back through allied forms in a closely connected series to early Tertiary times. In several cases the series are so complete that there can be little doubt that the line of descent has been established. The evolution of the horse, for example, is to-day demonstrated by the specimens now known. The demonstration in one case stands for all. The evidence in favor of the genealogy of the horse now rests on the same foundation as the proof that any fossil bone once formed part of the skeleton of a living animal. A special creation of a single bone is as probable as the special creation of a single species. The method of the paleontologist in the investigation of the one is the method for the other. The only choice lies between *natural derivation* and *supernatural creation.*

For such reasons it is now regarded among the active workers in science as a waste of time to discuss the truth of evolution. The battle on this point has been fought and won.

The geographical distribution of animals and plants, as well as their migrations, has received much new light from paleontology. The fossils found in some natural divisions of the earth are related so closely to the forms now living there, that a genetic connection between them can hardly be doubted. The extinct Marsupials of Australia and the Edentates of South America are well-known examples. The Pliocene hippopotami of Asia and the south of Europe point directly to migrations from Africa. Other similar examples are numerous. The fossil plants of the Arctic region prove the existence of a climate there far milder than at present, and recent researches at least render more probable the suggestion, made long ago by Buff on, in his "Epochs of Nature," that life began in the polar regions, and by successive migrations from them the continents were peopled.

The great services which comparative anatomy rendered to paleontology at the hands of Cuvier, Agassiz, Owen, and others have been amply repaid. The solution of some of the most difficult problems in anatomy has received scarcely less aid from the extinct forms discovered than from embryology; and the two lines of research supplement each other. Our present knowledge of the vertebrate skull, the limb-arches, and the limbs, has been much enlarged by researches in paleontology. On the other hand, the recent labors of Gegenbaur, Huxley, Parker, Balfour, and Thacher will make clear many obscure points in ancient life.

One of the important results of recent paleontological research is the law of brain-growth, found to exist among extinct mammals, and to some extent in other vertebrates. According to this law, as I have briefly stated it elsewhere, "all Tertiary mammals had small brains. There was, also, a gradual increase in the size of the brain during this period. This increase was confined mainly to the cerebral hemispheres, or higher portions of the brain. In some groups, the convolutions of the brain have gradually become more complicated. In some, the cerebellum and the olfactory lobes have even diminished in size." More recent researches render it probable that the same general law of brain-growth holds good for birds and reptiles from the Mesozoic to the present time. The Cretaceous birds, that have been investigated with reference to this point, had brains only about one third as large in proportion as those nearest allied among living species. The Dinosaurs from our Western Jurassic follow the same law, and had brain-cavities vastly smaller than any existing reptiles. Many other facts point in the same direction, and indicate that the general law will hold good for all extinct vertebrates.

Paleontology has rendered great service to the more recent science of archæology. At the beginning of the present period, a reëxamination of the evidence in regard to the antiquity of the human race was going on, and important results were soon attained. Evidence in favor of the presence of man on the earth at a period far earlier than the accepted chronology of six thousand years would imply, had been gradually accumulating, but had been rejected from time to time by the highest authorities. In 1823, Cuvier, Brongniart, and Buckland, and later, Lyell, refused to admit that human relics, and the bones of extinct animals found with them, were of the same geological age, although experienced geologists, such as Boué and others, had been convinced by collecting them. Christol, Serres, and Tournal, in France, and Schmerling in Belgium, had found human remains in caves, associated closely with those of various extinct mammals, and other similar facts were on record.

Boucher de Perthes, in 1841, began to collect stone implements in the gravels of the valley of the Somme, and in 1847 published the first volume of his "Antiquités Celtiques." In this work he described the specimens he had found, and asserted their great antiquity. The facts as presented, however, were not generally accepted. Twelve years later. Falconer, Evans, and Prestwich examined the same localities with care, became convinced, and the results were published in 1859 and 1860. About the same time, Gaudry, Hébert, and Desroyers also explored this valley, and announced that the stone implements there were as ancient as the mammoth and rhinoceros found with them. Explorations in the Swiss lakes and in the Danish shell heaps added new testimony bearing in the same direction. In 1863 appeared Lyell's work on the *"Geological Evidences of the Antiquity of Man,"* in which facts were brought together from various parts of the world, proving beyond question the great age of the human race.

The additional proof since brought to light has been extensive, and is still rapidly increasing. The Quaternary age of man is now generally accepted. Attempts have recently been made to approximate in years the time of man's first appearance on the earth. One high authority has estimated the antiquity of man merely to the last glacial epoch of

Europe as 200,000 years; and those best qualified to judge would, I think, regard this as a fair estimate.

Important evidence has likewise been adduced of man's existence in the Tertiary, both in Europe and America. The evidence to-day is in favor of the presence of man in the Pliocene of this country. The proof offered on this point by Professor J. D. Whitney in his recent work is so strong, and his careful, conscientious method of investigation so well known, that his conclusions seem irresistible. Whether the Pliocene strata he has explored so fully on the Pacific coast corresponds strictly with the deposits which bear this name in Europe, may be a question requiring further consideration. At present, the known facts indicate that the American beds containing human remains and works of man are as old as the Pliocene of Europe. The existence of man in the Tertiary period seems now fairly established.

In looking back over the history of paleontology, much seems to have been accomplished; and yet the work has but just begun. A small fraction only of the earth's surface has been examined, and two large continents are waiting to be explored. The "imperfection of the geological record," so often cited by friends and foes, still remains, although much improved; but the future is full of promise. In filling out this record, America, I believe, will do her full share, and thus aid in the solution of the great problems now before us.

I have endeavored to define clearly the different periods in the history of paleontology. If I may venture, in conclusion, to characterize the present period in all departments of science, its main feature would be a *belief in universal laws*. The reign of Law, first recognized in the physical world, has now been extended to Life as well. In return, Life has given to inanimate Nature the key to her profounder mysteries—Evolution, which embraces the universe.

What is to be the main characteristic of the next period? No one now can tell. But, if we are permitted to continue in imagination the rapidly converging lines of research pursued to-day, they seem to meet at the point where organic and inorganic nature become one. That this point will yet be reached, I cannot doubt.